Gustavo Pereira Schier Junior
Zoraide Fonseca Costa

Environmental Sustainability

Gustavo Pereira Schier Junior
Zoraide Fonseca Costa

Environmental Sustainability

and the Disposal of Automotive Lubricating Oils in Guarapuava-Pr

ScienciaScripts

Imprint

Cover image: www.ingimage.com

This book is a translation from the original published under ISBN 978-3-330-75495-9.

Publisher:
Sciencia Scripts
is a trademark of
Dodo Books Indian Ocean Ltd. and OmniScriptum S.R.L publishing group

120 High Road, East Finchley, London, N2 9ED, United Kingdom
Str. Armeneasca 28/1, office 1, Chisinau MD-2012, Republic of Moldova, Europe
Printed at: see last page
ISBN: 978-620-8-26919-7

SUMMARY

DEDICATORY

I dedicate this work to my daughter Ana Clara Schier, I do everything for her, she is the reason I improve every day as a human being, and to everyone who believed that it would be possible to complete the course.

ACKNOWLEDGMENTS

I would first like to thank the Lord of Hosts, who gave me the strength and courage to complete this degree, and especially:

My beloved wife Daiana, my "asset", who mentored and encouraged me to go to university and complete my degree.

To my family, my sister Telma, my niece Marina, my nephew Giovanni and my brother-in-law, whose help was essential in completing the course.

To Professor Zoraide who helped me, supported me at various times and played a crucial role in this work.

To all the teachers at Decon, who dedicatedly passed on their knowledge without measuring their efforts.

To my classmates, especially Anderson Mendes and Erllandson Lopes, who over the four years have been essential to my growth as an academic and as a person.

SUMMARY

Environmental sustainability is a subject that has sparked research and discussion among economists, environmentalists, academics and other professionals involved in the field. Inserted in this context is the disposal of automotive lubricating oils, an item that has a life cycle that can be reused countless times depending on how it is managed. The aim of this work is to investigate how automotive lubricants are disposed of in the city of Guarapuava-Pr, as well as to analyze the work of reverse logistics and compliance with specific recycling laws. Methodologically, a quantitative study was carried out in which the data was tabulated using tables, graphs and percentages. The sources of the data collected were books on environmental economics, articles and scientific journals, and finally a case study with interviews with the actors involved in the work. The results showed that there is efficiency in the geographical scope served by the companies that carry out the Reverse Route for Used or Contaminated Lubricating Oils in the Brazilian regions, with good results in the South and especially in the state of Paranâ. With regard to the city of Guarapuava-Pr, it was found that environmental laws are being applied in some of the establishments surveyed. However, in other places it was found that there is no knowledge of current recycling laws and there is also no environmental awareness on the part of those who use lubricating oils as a means of business or related economic activity. Therefore, there is a need to follow and monitor the correct disposal of automotive lubricating oils in order to inhibit the environmental impacts caused by these products.

Key words: Sustainability, Oluc, Reverse Logistics, Environment.

I. INTRODUCTION

Lubricating oils, motor oils or engine oils are substances that work with moving parts or components that use a lubricating fluid to reduce friction between the parts and prevent wear of their internal and moving parts, most often a mineral oil formulated from petroleum.

Lubricating oil must be changed constantly in motor vehicles, allowing the vehicle to function efficiently. Failure in the lubrication system can result from incorrect use of the lubricant recommended by the manufacturer, causing technical and mechanical problems such as the formation of sludge and strange noises, overheating, increased fuel consumption and loss of power.

The collection of Used or Contaminated Lubricating Oil (Oluc) must be carried out by a specialized company, carrying out Reverse Logistics so that the re-refining process can take place, which consists of processing the product, making its life cycle essential, until the product obtains the characteristics of base oils, for the formulation of new products. In addition to environmental commitment, this research aims to address some economic issues, such as environmental economics and product supply and demand.

The focus of this work will be on the collection points for oil and the company that collects the discarded products. In establishments and fleet companies where oil is changed, there must be proper packaging for storing the discarded oil so that it can be collected correctly. This discarded oil is taken to the re-refining industry.

This contamination usually causes pollution, which damages the health of the workers who handle it and the people who come into contact with the product. Contamination of air, water, soil and food also causes a number of serious problems, such as respiratory problems, carcinogens and adverse effects on reproduction.

The issue addressed in this work is whether there is effective monitoring by the government agency responsible for the correct disposal of Oluc in the city of Guarapuava-PR. This is associated with the awareness of the players who use these products in their production and commercial chain, together with the obstacles caused by the incorrect disposal of Oluc, which is a problem that affects all citizens. By damaging nature, the health of the population is compromised, as well as everything related to it. In addition to health, this improper disposal can lead to illegal trade in these products, diverting them from the reverse route, causing environmental impacts, preventing tax collection and generating jobs and income.

This is presented as a hypothesis in the light of the currents and the problem cited for this research. It is assumed that as Guarapuava is one of the largest cities in the state of Paraná, the inspection and collection of Olucs should occur normally without major adversities. Olucs are products that degrade slowly in nature, causing various health problems, such as respiratory organs and chronic obstructive pulmonary diseases, as well as diverting the environmental focus and causing various harm to fauna and flora. If Oluc is disposed of incorrectly, the illegal trade in these products can result, leading to the detour of these products from re-refining and reverse logistics, causing serious obstacles that can damage the environment, leading to violations of the law on recycled products and environmental crimes.

The aim of this work is therefore to investigate whether the disposal of automotive lubricating oils is being carried out correctly in general in the city of Guarapuava-Pr, as well as the knowledge and awareness of the players in this production chain about recycling laws and the harm caused to the environment.

The specific objectives will be: to demonstrate the geographical scope served by the collection of lubricating oils, to know the potential and behavior of the automotive lubricating oil market in the city of Guarapuava - PR, to identify ways that can be useful to avoid the incorrect disposal of lubricating oils in nature.

The aim of this work is to demonstrate a four-chapter structure, plus the introduction. The conceptual theoretical foundation is presented in Chapter II, which is subdivided into three sections. The first presents a theory of Environmental Economics, highlighting its natural resources and policies. The second section will emphasize reverse logistics, as well as its importance in reducing environmental impact and sustainable development, the environmental impact caused by the product studied and its externalities, a field of microeconomics. The third section will give a history of Olucs, their reverse logistics and national legislation. The methodology applied will be described in Chapter III, setting out the characterization of the research, bibliographic sources and data collection strategy. The following chapters IV and V present the results and discussions and the final considerations.

II. CONCEPTUAL THEORETICAL FOUNDATION

In order to achieve the objective of this chapter, it will be presented how Environmental Economics is viewed by some authors, its policies and characteristics over time, as for Reverse Logistics, the publications related to the area of natural and agricultural resources are pertinent in the course of the proposed theme. And finally, how the practical part is subject to theory, as well as current research legislation.

2.1 Environmental Economics

The study of the economics of natural resources has acquired great notoriety in various currents of thought, but the dominant and contentious approach is with neoclassical economics, or conventional economics. It is therefore necessary to understand its proposals and adversities, and to look at some mathematical instruments for the resources imposed and the modeling chosen (MAY, 2010).

According to May (2010), this current is not absolute; it includes other broader and more contemporary approaches that are a way of expanding this field of study, such as ecological economics. In this way, approaches to other approaches should be made in order to counter a dominant and critical view.

At the beginning of the formation of economic theories, renewable natural resources played a role:

> explanation of the source of material wealth. This is expressed 1) in the physiocratic theses, which in the second half of the 18th century claimed that the agricultural sector was the source of all surplus, 2) in the warning of the classical school, at the beginning of the 19th century, about the possible jeopardization of capitalist expansion as a result of the scarcity of natural resources, perceived by the imbalance between population growth and food supply, according to Thomas Malthus, and by the reduction in the productivity of agricultural labor - due to the scarcity of fertile land - and the consequent fall in profits, in the famous "land rent theory" enshrined by David Ricardo; 3) in addition to theses such as Jevons', from the second half of the 19th century, which highlighted great concern about the indiscriminate use of coal in England, which would lead to the exhaustion of this resource, so vital to the country's development process (MAY, 2010, p. 49).49).

In this way, the author tries to link environmental economics to neoclassical economics, using already solid theories from economists who have made a mark in economic history, because there are many concerns about environmental adaptation and sustainability projects in the debates between environmentalists and economists, but in order for this to happen, there has to be collaboration so that environmental economics comes out of theory and into practice.

According to Moura (2006), the environmental approach takes into account the fact that natural resources are limited, so their use must be done in a sustainable way, giving rise to the need to work with the scarcity of available resources. The economic variable is always present in this interaction, since the implementation of new laws, the demands and pressures of consumers and the very awareness of entrepreneurs are factors that force a new attitude and new rules of conduct with regard to industrial activities.

With the growth of discussions on sustainable development, it is clear that there is a need for a differentiated analysis in this branch of economics, since it deals with both economic and cultural performance, and that the choice of entrepreneurs and society must be assertive with regard to environmental decisions in the formation or restructuring of a company.

According to Loyola (1997), environmental economics goes hand in hand with the other exact science, physics, as each seeks its own balance, while economics appears as a science based on mechanical physics. In this way, Adam Smith's idea of the invisible hand is nothing more than an assimilation borrowed from the natural order of things established by physics.

Just as a body also tends towards equilibrium if there is no force to alter it, for the economy this situation of equilibrium is possible because the different economic agents, in pursuit of their personal interests, lead to this situation of equilibrium, thus bringing resolutions and even economic forecasts to society (LOYOLA, 1997).

A new discussion of environmental problems that has been widely analyzed is the use of natural resources. As mentioned earlier in the study, for environmental economists there are no problems of absolute scarcity, only relative scarcity (LOYOLA, 1997).

Therefore, from this point of view, it is possible that certain types of resources may be temporarily exhausted. Their proposal is that given the existence of renewable and non-renewable resources, through technological development it is possible to replace renewable resources with non-renewable ones.

But as some authors have pointed out, it is not always possible to make this substitution, since natural resources have their own characteristics that cannot always be reproduced by other sources:

> in the use of non-renewable resources puts a society's economic development at risk. It should be mentioned that this confidence in technological development is due to the fact that in many cases it can be proven that new technologies allow for an increasingly efficient use of resources, making their use less (LOYOLA, 1997, s/p).

The author addresses an empirical form of the application of environmental economics in society, a fact that requires time for new experiments and plausible conclusions so that investment in new technologies applied to non-renewable resources can proceed.

In environmental economics, the solution to the problem of a lack of resources is only seen as a technological problem. If technology develops to the levels required so that renewable resources can be exchanged for non-renewable ones, this will make scarce resources useful again. As this is unfeasible, then:

> technological changes that seek greater resource efficiency should be achieved. Of course, this is only possible if we assume that the problem is one of transformation. The logic proposed is correct if we believe that the environmental problem can be summed up in questions involving only the law of conservation of matter (LOYOLA, 1997, s/p).

Therefore, the proposed way of inserting environmental economics into applied economics will be subject to technological changes as well as changes in thinking, because when the conclusion is reached that so much technological investment in non-renewable resources is unfeasible.

Traditional scientific economics, in fact, does not consider any connections that may exist between the ecological system and the activities of producing and consuming that represent the core of any economic system (activity-economy). The typical economic model does not take into account the environmental framework or constraints. It only focuses on flows and variables in the economic domain, and does not include opportunity costs, among other attributes that traditional economics applies to the environment (CAVALCANTI, 2010).

In classical economics, the environmental sector is considered to be an externality according to Samuelson (1967) *apud* Cavalcanti (2010). It does not fit into the monetary flow of income where producers and consumers meet, because in this model the economy is self-sustaining and has no barriers to overcome:

> orthodox economics treats environmental impacts, if it deals with them at all, as phenomena external to the economic system, seen as market failures. For it, externalities can, with appropriate methods, be internalized into the price system: a way, it supposes, of correcting market failures (CAVALCANTI, 2010).

However, the authors have a similar line of thought, as they both point out the difficulty of including environmental economics within pure economics, due to the high technological investment, as well as in studies of non-renewable resources, due to the fact that there is a period of scarcity, thus making it unfeasible to develop assertive research.

2.2 Environmental policies and the classification of natural resources

Two environmental policies will be highlighted in this study: the "Green Protocol" and the

"Solid Waste Policy", where the first is a policy that generalizes companies in order to promote environmental adjustments and the second policy is based on the product studied in this research.

Environmental policy consists of a set of measures and targets aimed at reducing the negative impacts on nature caused by man's actions. In order to make sense of its existence, it has theories, targets, instruments and provides for penalties and fines imposed by the responsible environmental body, as well as supervision for those who do not comply with these measures, and funding for various companies that seek to comply with environmental legislation (MAY, 2010).

Still contemplating what the author says, we see that the importance of environmental policy has been growing, especially in industrialized and developed countries, as its effects on international trade can be seen: "With the emergence of non-tariff barriers, as each country has specific environmental problems, there are differences in the principles and types of environmental policy instruments adopted, but there are general features that are common to all countries" (MAY, 2010, p.163).

In Brazil, the federal government launched a program called the "Green Protocol" in November 1995, which stipulates that companies wishing to obtain financing from federal banks must carry out environmental risk analyses on their projects and demonstrate that the risks are acceptable. On that occasion, the following banks signed the letter of commitment to sustainable development: Banco do Brasil, Caixa Econômica Federal, Banco da Amazônia, Banco do Nordeste, Banco Nacional de Desenvolvimento Econômico e Social and Banco Central. In this way, the government has created credit lines and incentives for companies that do not emit effluents into the environment (MOURA, 2006).

Later, in 2010, the "Solid Waste Policy" was instituted, which became Federal Law 12.305/2010, where Solid Waste refers to pollutants resulting from the incomplete burning of oil left over from engines, an important aspect of the principles is to achieve prevention and precaution, sustainable development, cooperation between different areas of the public sector and society in general, linked to the goal of greater environmental education, protection of public health, recycling and treatment of solid waste as well as proper final disposal of waste (BRASIL, 2010).

According to this policy, there are a number of economic instruments aimed at providing funding to private institutions for research and development projects that boost the development of solid waste management, sustainable projects, reverse logistics, aimed at improving production processes and the reuse of waste (BRASIL, 2010).

According to the Brazilian Waste Treatment Association (Abetre), there is a regulatory standard for such waste, the number of which is: 10.004:2004, and it also contains some aspects regarding its classification, which involve identifying the process or activity that gave rise to it, its constituents and characteristics, and comparing these constituents with lists of waste and substances whose impact on health and the environment is known (ABETRE, 2006).

It is important to have an appropriate environmental policy that can be applied to society in general, where those who use natural resources are not harmed by a lack of information and ignorance of how to apply these resources. Where there can be broad knowledge of laws and tools that can be linked to other policies.

"The ability of a renewable natural resource to replenish itself over the horizon of human time has been something of a challenge, and one of the criteria for classifying resources, which can be renewable, or reproducible, and non-renewable" (MAY, 2010, p.50), also known as exhaustible, which is the product in focus in this research. There are some examples of natural resources in Brazil, as shown in Table 1.

Table 1: Renewable natural resources in Brazil

Biomes	Environmentalists claim that the Cerrado will disappear by 2030.
Soils	In northeastern Brazil, land use is being jeopardized by the increasing rate of desertification, which is growing every year.
Timber forest resources	According to the Food and Agriculture Organization of the United Nations (FAO), Brazil has the worst deforestation record in the Amazon.
Water	According to information from the Center for Strategic Studies and Management (CGEE), the total amount of water withdrawn globally from rivers, aquifers and other sources has increased ninefold, while use per person has doubled and the population has grown threefold.
Exhaustible resources	Discoveries of oil reserves will make Brazil one of the world's top ten reserves by the end of the 2010s.
Non-renewable resources	When steel is produced from scrap, it saves 70% of what is spent on producing the original ore.
Aluminum recycling	It can generate any product several times, without

	losing quality in its reuse process.

Source: MAY, 2010.

However, as oil is an exhaustible commodity, and lubricating oils are formulated by removing base oils, we see that the outlook for the manufacture of lubricating oils does not tend to diminish or become a scarce product, as long as it is used correctly and the competent bodies take due care with the oil deposits recently discovered in Brazil.

2.3 Externalities and Environmental Impact

Externalities can occur between the actions of producers and consumers, they can be negative or positive, depending on the case and how they occur. Negative externalities occur when one of the actions imposes costs on the other, thus causing marginal external costs and social costs (PINDYCK, 1994).

When it comes to automotive lubricants, it is common for accidents to occur in pipelines that transport these products by sea, as it is a cost-competitive form of transportation and a load with a greater storage capacity than other means of transport can handle. As a result, large leaks can occur, resulting in a major environmental disaster, causing many problems for those who use the means of survival, such as fishermen, surfers, as well as the fauna that is harmed, causing a great social cost.

Externalities can have a negative or positive effect, but they are perfectly symmetrical, so they are treated as an external economy for a positive effect or an external diseconomy for a negative effect. For Pigou (1920), the economy or diseconomy can be analyzed in terms of the divergence between private cost and social cost, where social cost is taken to mean the cost for all the economic agents that make up the community (FAUCHEUX, 1995).

According to the author, in the welfare economics of Pigou's time, if a company's production entails any kind of harm that directly affects economic agents and is not compensated for by the market, the social marginal cost of production is higher than the private marginal cost of production.

When there is a gap between the social cost and the private cost, this is called internalization, the method proposed by Pigou of which is to fill the gaps in these costs by paying a fee to the emitter of the damage, a fee represented by figure 1, the difference between the social cost and the private cost.

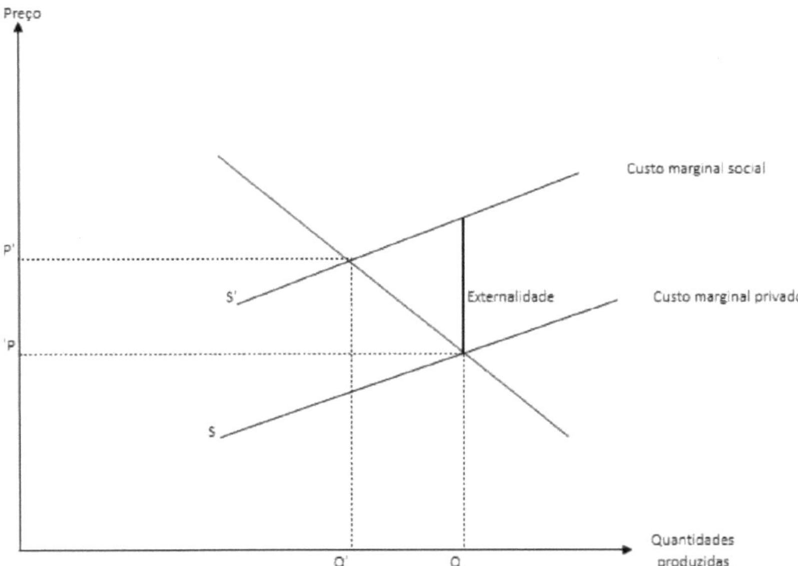

Figure 1: Rigorous Externality, Distance Between Social Cost and Private Cost Source: FACHEUX (1995).

When it comes to automotive lubricants, it is common for accidents to occur in pipelines that transport these products by sea, as it is a cost-competitive form of transportation and a load with a greater storage capacity than other means of transport can handle. As a result, large leaks can occur, resulting in a major environmental disaster, causing many problems for those who use it as a means of survival, for example: fishermen, surfers, in addition to the fauna that is harmed, causing a great social cost.

Lubricating oils, engine oils or motor oils are substances that work with moving parts or components that use a lubricating fluid to reduce friction between the parts and prevent wear of their internal and moving parts, most often a mineral oil formulated from petroleum.

Because it is not biodegradable, used or contaminated lubricating oil takes dozens of years to disappear from the environment and, when disposed of improperly, causes great damage. For example, one liter of used or contaminated lubricating oil can contaminate one million liters of water, contaminate 1,000 m^2 of water surface (it forms a toxic surface layer that hinders the passage of light and oxygen exchange, causes a drop in aquatic photosynthesis and the death of fauna and flora). Contaminated water is easily absorbed by organisms and kills everything around it. The combustion of one liter of lubricating oil releases one gram of heavy metals into the air[1] . It totally renders the affected soil unusable, both for agriculture and for building, causing serious health problems and also for the economic development of regions affected by such carelessness on the part

1 The maximum acceptable emission of heavy metals is 5 mg/m3(N). 1.5 mg/m3 of lead is considered an excessive level.

of the players involved in this chain (SOHN, 2006).

This contamination generally causes surface water pollution; it impacts groundwater and aquifers (future drinking water reserves). Because it contains various elements that are harmful to health, such as chromium, cadmium, lead and arsenic, it causes damage to the health of the workers who handle it and the people who come into contact with it. Contamination of the air, water, soil and food also causes a number of serious problems, such as respiratory problems, carcinogens and adverse effects on reproduction and the development of the fetus (SOHN, 2006).

One of the main characteristics of Oluc is that there is no prospect of reducing its generation, even though this product is characterized as exhaustible. The more technology there is, the more contamination there can be, and this is one of the sectors that is developing the most along with the growth of the car fleet in Brazil.

Large oil spills cause serious problems for the community, especially for those who still depend on fishing, and also for the fauna that lives in the open sea.

An oil spill can occur in many ways, causing enormous damage and harming the environment, but the biggest events usually involve a pipeline rupture at sea or a well blowout. The causes of pipeline ruptures are diverse and include damaged pumping equipment or lack of necessary preventive maintenance, earthquakes, tsunamis, sabotage, deliberate oil spills, among other events. Because of the extensive use of sensors and mechanisms to interrupt sections of pipelines, accidental events are much smaller than those that occur in ocean tanks or off-shore platform explosions (SILVA, 2009).

Following the same concept as Silva (2009), operational discharges from oil tank washes are also frequent sources of marine spills. Such operations in the open sea are forbidden, but it is natural for them to continue to be abused, given the difficulties in monitoring them, and they are normally used by large companies that make a relatively satisfactory profit from importing oil derivatives. It is important to know that the size of a spill does not necessarily indicate its potential to cause damage, as a small accident can lead to serious damage to a highly sensitive environment.

Consequently, the type of oil product can affect the severity of the ecological damage. The most important considerations are the degree of toxicity and the environmental persistence of the spilled materials, which means that it takes time and a high level of investment to research and respond to the extent to which the disaster has affected the environment (SILVA, 2009).

In view of this, the state should intervene more with oil importing companies with regard to logistics and the maintenance of equipment that transfers these environmentally harmful products, preventing social damage regardless of the location of the region and the economic activity it carries out.

2.4 A Brief History of Lubricating Oil Recycling

In the 1960s, the National Petroleum Council (CNP) regulated the re-refining activity in Brazil, due to the government's interest in reducing oil imports, a source of raw material for the production of lubricating oils, The reduction in imports at the time would have meant a reduction in supply, to be met by the recycling of used oils. Since then, various laws have been introduced to regulate the sources that generate used oils, their collection and the quality standards for lubricants (LEITE, 2009).

According to the National Development Bank (BNDES), the demand for lubricants in Brazil has led to an increase in imports in this segment. From 2008 to 2012, there was a 30% increase in the volume of imports, bringing the segment's deficit to almost 1 billion dollars in 2012. Base oils were responsible for approximately 60% of this deficit, while lubricants accounted for 23% and additives 17% (BNDES, 2014).

It can be seen that the balance of trade is unfavorable when it comes to lubricating oils, because they are products that can be extracted from base oil with greater use of Arab oil, which makes it possible to get a higher yield from the product. More research and development is needed so that Brazilian oil becomes more competitive and a greater quantity of base oil can be extracted for mixing with the finished lubricant.

Still, according to Leite (2009), until 1988, the legislation provided for a tax to encourage the reverse logistics of waste oils, by means of a single tax, which became economically attractive and would be better than any inspection for the generators of the oils and the points of sale of the product, The amounts collected from the sale of Olucs were used to help the company's cash flow, investments in marketing and even to make up the payroll, but in the same year the government decreed the end of the single tax, causing a sudden lack of interest in the recycling operation for the reverse chain.

Following this same thought, it was noted that this unfeasibility resulted in the closure of several re-refining companies, due to the increase in transportation and consolidation costs, going from 32 companies in Brazil to 8 companies, where its impact was greater outside the Rio Sâo Paulo axis, due to the re-refining companies being located in the Brazilian Southeast, leaving the rest of the region of the federation with little coverage by reverse logistics, there was a drop in the amount of Oluc in the sector from 250 million liters to 140 million liters per year.

At the beginning of the 1990s, the government granted a tax subsidy to the sector, called the Uniform Price Fund (FUP), with the aim of improving the relationship between the flows of used and contaminated oils in the regions furthest from the Southeast. In 1997, new legislation eliminated the FUP from the Olucs, and assigned responsibility for organizing reverse logistics and industrial recycling to the companies that produced new lubricants.

This standardization provided for an initial period until the end of 2000, after which the responsibility for the entire Oluc reverse chain would fall to the producers. This regulation provided for recycling rates, starting with 20% in the first year, 25% in the second year and 30% in the third year.

2.5 Peculiarities of re-refining

According to Tristao (2005), lubricating oils account for around 2% of all petroleum products and are the only ones that can be kept with their original characteristics, performing their functions for periods as long as 20 years or more in sealed units, exercising the re-refining chain, where the product will return to its original form.

Lubricants lose their ability to perform their function due to the presence of internal or external contaminants, called solid residues, as well as the loss of action of additives. However, when subjected to appropriate regeneration treatments, such as re-refining, they return to their original form. Olucs are processed when they are transformed into base oil to be returned to the industry for the manufacture of new oil, thus completing their life cycle, as lubricants lose their ability to perform their function due to the presence of internal or external contaminants, as well as the depletion of additives (TRISTÂO, 2005).

According to the same author, used oil is a resource that, if properly recycled, can be returned to the production chain unlimited times, suffering only the losses of each process, generating considerable economic advantages and saving foreign currency in the case of countries such as Brazil.

However, when subjected to appropriate restructuring treatments, such as re-refining, Oluc returns to its normal form, ready to be reused and marketed between retailers and generators, thus boosting regional economies along the product's production chain. As illustrated in figure 2, the re-refining process looks like this:

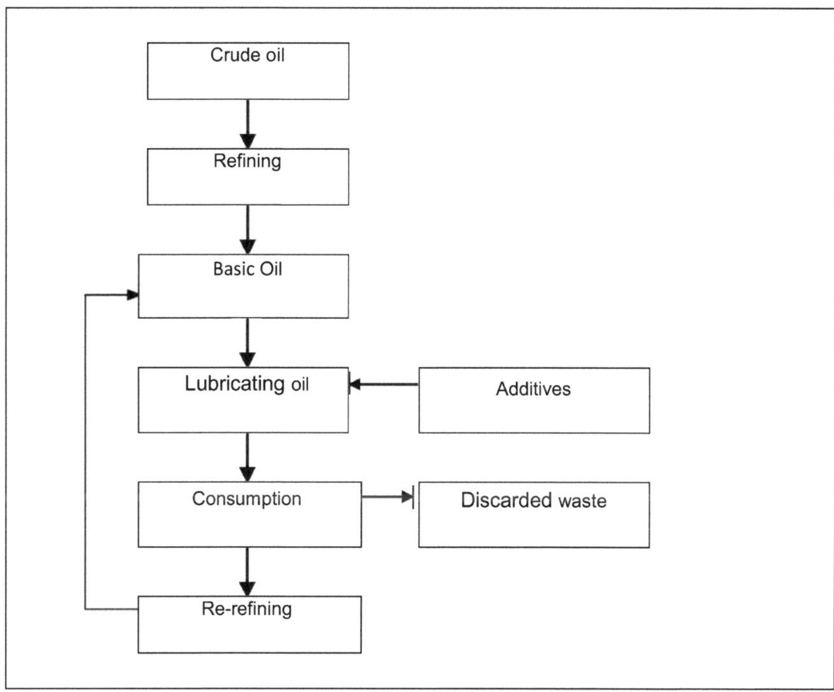

Figure 2: Re-refining cycle

Source: TRISTÂO (2005).

Therefore, it can be seen how economically viable re-refining is, as it inhibits the country from importing larger quantities of lubricating oils from Arab oil, as well as containing the environmental impact caused by these products. Re-refining has fundamental value in the Olucs production chain, as it makes this item sustainable.

The work of Gonçalves (2014) lays out the stages of the re-refining process for used lubricating oils, from their collection, processing and management to where the solid waste should be sent. The stages are shown in Table 2 below:

Table 2: The re-refining process

Stages	Description
A) 1st reception and filtration	The used lubricating oil is unloaded and homogenized. It is then analyzed by the company's Quality Control department in accordance with the guidelines

	established by the ABNT standards and, once the analysis has been approved, the product is filtered and stored in appropriate tanks, located in containment basins.
B) 2nd dehydration	At this stage of the process, the used lubricating oil is heated up to 120°C to remove the water, and up to 280°C to remove the organic compounds with low molecular weight carbon chains. The system is equipped with a series of heat exchangers, which make energy use of the heat generated and the fractions that require thermal exchange.
C) 3° Total Evaporation	The dry lubricating oil from the dehydration process is sent to the total evaporation unit. The process consists of applying high temperature (above 375 °C), high vacuum and centrifugal force to separate the heavier fractions contained in the oil. These fractions are separated by evaporation and then condensed again using powerful condensers.
D) 4th Physico-Chemical Treatment	The oil coming from the total evaporation unit, already cooled to room temperature, still has some quantities of oxidized compounds to be separated. To extract them, a flocculating agent is applied, in minute quantities, which promotes the agglomeration of the oxidized compounds, which then decant and are separated after a few hours.
E) 5° Clarification	The lubricating oil from the physical-chemical treatment is pumped to the clarification system, where it receives the addition of a clarifying agent. This process is responsible for absorbing the particles that give the oil its color. The temperature is around 350 °C, and steam is used to drag out the light fractions that may still be present in the oil. To guarantee the quality of the re-refined oil, laboratory analyses are carried out at this stage to meet the Quality Parameters.
F) 6th filtration	The oil mixed with the clarifying agent passes through a system of filter presses and bags to remove the particulates. It is then pumped into the base oil tanks and re-refined at room temperature. This re-refined oil meets the highest demands of a mineral base oil.
G) 7th Waste Management	All industrial waste generated by the industry can be sent to cement industry kiln co-processing units, duly

	licensed by the competent environmental bodies. Likewise, industrial effluents must be treated in accordance with the relevant guidelines

Source: GONÇALVES (2014, s/p)

The re-refining process has a positive track record in terms of sustainability, where products are reused and returned to their original form several times, thus benefiting the environment and society, and can also slow down the process of importing this product.

2.6 Irregular disposal of Oluc

The guidelines issued by the National Environment Council show that there is a real problem with the incorrect disposal of waste oil, causing socio-economic problems for the society that uses these products for consumption and also for commerce. Despite the obvious risk it poses to the environment, Oluc is a very important source of the essential raw material that is basic lubricating oil, necessary for the formulation of finished lubricating oil, which is only found in significant quantities in imported light Arab oil (CONAMA, 2005).

This body states that the management of Oluc must be appropriate, as recommended by the use of this waste, which is not only environmentally important, but also economically important, and is even of great importance in the strategy of national self-sufficiency in relation to oil.

Another particularity of Oluc, which is included in these guidelines, is that, unlike other waste, it has an attractive economic value for purposes other than its correct, legally established environmental destination, making it a product that has a good return and is traded illegally.

According to the same body's rules, the practice to be curbed in this context is the irregular shipment of lubricating oil for illegal uses, such as: use as a fuel which is expressly prohibited, adulteration of fuels or lubricating oils, paint composition, grease formulations, so-called "traditional" uses.

Although all these detour are more a matter for environmental inspection than licensing, it is important for the environmental officer to be alert to possible detour attempts as early as the assessment of the project's proposed structure, and to be able to contain illegal trade or incorrect disposal in the environment (CONAMA, 2005).

A classic example that the same Council points out is the use of Oluc, which is considered by these standards to be the fuel composition of companies that own fleets of cars or trucks, or its adulteration into grease or waterproofing. The same happens with companies that propose a revolutionary recycling method, but through totally inadequate processes.

However, it is clear how much the correct disposal of waste oil means economically, because as well as complying with environmental laws, it will generate products and jobs both in the logistics area and in other sectors of the companies responsible for collecting and re-refining the oil collected.

According to CONAMA (2005), the environmental licensing guidelines for the recycling of Olucs list the following players and their responsibilities, as shown in the table below.

Table 3: Actors in the Olucs production chain

1. Theproducers and importers	They are legal entities that introduce the finished oil to the market and have a legal obligation to pay for its collection and to inform consumers (generators) of their obligations and the environmental risks arising from any illegal disposal of the waste.
2. Resellers	These are legal entities that sell finished lubricating oil at wholesale and retail, which, among other obligations, must receive the Oluc from the generators, in suitable facilities.
3. The generators	These are physical or legal persons who, as a result of using lubricants, generate Oluc, and who are obliged to deliver this hazardous waste to the collection point (dealer) or authorized collector.
4. The collectors	They are legal entities duly licensed by the competent environmental agency and authorized by the oil industry regulator to carry out the activity of collecting Oluc and delivering it to the re-refiner.
5. The re-refiners	Re-refiners are legal entities duly authorized by the oil industry regulator and licensed by the competent environmental agency for re-refining activities, whose obligation is to remove contaminants from hazardous waste and produce basic lubricating oil in accordance with ANP specifications.

Source: CONAMA (2005).

In this way, the players in the chain involved mainly in the marketing of lubricating oils must always act in accordance with the regulations of Conama 362/2005, following the rules so that there is no degradation of nature.

2.7 Reverse Logistics and Sustainability

Reverse Logistics plays a fundamental role in the life cycle of Oluc, because it is through this means that there is collaboration for its improvement and the reduction of environmental impact, being an important agent for the management of these products, because through Reverse Logistics the regional economy can be heated up, generating jobs and tax collection for the State.

After the Second World War, reverse logistics became an essential part of business life with the emergence of computers, as it became a support for new production technologies in industrial companies. In the high-speed response scenario, the logistical location of supplier companies, the control and transportation of supplies of components with high frequency and in small quantities, the long-term purchase and sale contract become imperative, thus business logistics becomes vital to the success of companies that adopt such procedures for reverse logistics, as this demonstrates planning and good management (LEITE, 2009).

According to Leite (2009), there is little interest in the study of reverse distribution channels, because they are not economically attractive when compared to direct distribution channels, but reverse logistics is necessary because the volume transacted by this logistics is only a fraction of the distribution, so the value is relatively low compared to the original goods.

In the case of finished Lubricating Oils, according to the same author, there is a real need for Reverse Logistics, as it is a post-consumption product with strategic legal, economic and ecological objectives at the same time. In this situation, Reverse Logistics is encouraged because the product is considered harmful to health and needs to have specific legislation.

According to Sindirrefino (2012), Reverse Logistics is an instrument of economic and social development characterized by a set of actions, procedures and means whose purpose is to enable the collection and return of solid waste to the business sector for reuse in its cycle or in other production cycles.

Figure 3 shows the Reverse Logistics process from its starting point, which is Industry, through the agents in the chain until it completes its cycle:

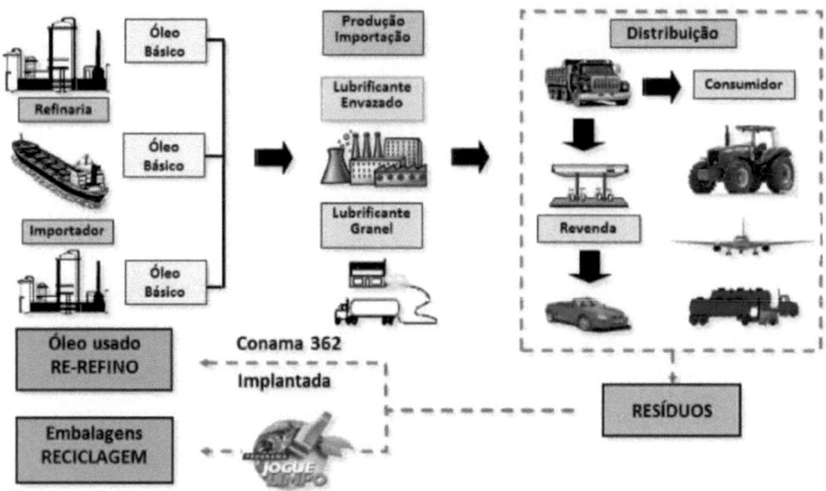

Figure 3: Reverse Logistics Cycle Source: Sindirrefino (2012).

By following this Reverse Route, the law will be complied with, so that there will be no problems with incorrect disposal of Oluc, environmental crimes and there will be what is expected inherent in sustainable development, while it becomes a positive mechanism that is economically emphasized.

A lot has been said about these words, as they are of great importance in the economic and environmental spheres, because there is a need for good reverse logistics planning so that there is sustainable development capable of not extinguishing natural resources.

In Pereira's (2013) view, there are many challenges to sustainability, including making corporate units sustainable. In this regime of intense change, companies are seeking to stand out from the rest, bringing customized products to the system, thus seeking to serve markets that are becoming increasingly specific over time. However, the consumer mentality is also changing in this process, and in order to evolve their processes and competitiveness, companies must also adapt to these behavioral changes.

Sustainability should therefore be seen by corporations as a strategy rather than a business challenge. In this case, the author defines it as intelligent action:

> identifying the challenges imposed by society and the planet on companies and the market, seeking new business strategies to be present in a future that is desirable in some respects and inexorable in others, planning for risks and opportunities can lead a company to strategies that "take advantage" while costs are not affected by the changes that are yet to come (PEREIRA, 2013, p.152).

Reverse Logistics has a business-focused role with thoughts on returns, in order to add some kind of value or try to recover as much value as possible in a product that is on the margins of the market, so not every Reverse Logistics process is sustainable, however some of the processes of this reverse route contain sustainability assumptions in their prerogatives (PEREIRA, 2013).

However, Reverse Logistics is a correlated variable of sustainability, because both need to be side by side for those who use these goods and services to have a new vision, and understand how necessary it is to apply these variables for future generations.

2.8 Legislation regulating Olucs

At the federal level, there is the National Environment Council, which has passed Resolution 362/2005, laws with penalties. According to this regulation, rules have been laid down for the recycling of waste oil. The articles of this regulation that are most significant for this research are commented on below, along with some laws that penalize the incorrect disposal of Olucs.

According to Article 1 of Resolution 362/2005 of the National Environment Council, all used or contaminated lubricating oil must be collected and disposed of in such a way that it does not adversely affect the environment and allows for maximum recovery, i.e. that the product is reused in its full proportion and of the constituents contained therein, in the manner set out in this resolution described below:

Art. 3 - All used or contaminated lubricating oil collected must be recycled through the re-refining process.

§ Paragraph 1 - The recycling referred to in the caput may be carried out, at the discretion of the competent environmental body, by means of another technological process with proven environmental efficiency equivalent or superior to re-refining.

§ Paragraph 2 - The processing of used or contaminated lubricating oil will be allowed for the manufacture of products to be consumed exclusively by the respective industrial generators.

§ Paragraph 3 - Once it has been proven to the competent environmental body that the destination provided for in the heading and paragraph 1 of this article is unfeasible, any other use of used or contaminated lubricating oil will depend on environmental licensing.

§ Paragraph 4 - The processes used to recycle lubricating oil must be duly licensed by the competent environmental body (CONAMA, 2005,s/p).

This article specifies how important it is to re-refine lubricating oils so that they can be used throughout their entire life cycle, i.e. products that are not destined for recycling need a specific license for their use issued by the competent authority.

According to art. 5 of the same resolution, the producer, importer and retailer of finished lubricating oil, as well as the generator of used lubricating oil, are responsible for collecting used or contaminated lubricating oil, within the limits of the attributions provided for in this resolution, in which case it would be interesting to have specialized monitoring by a professional who is able to audit whether this agreement is really being complied with.

With regard to Article 6, producers and importers of finished lubricating oil must collect or ensure the collection and final disposal of used or contaminated lubricating oil, in accordance with this Resolution, in proportion to the total volume of finished lubricating oil they have sold, i.e. they cannot collect an insufficient quantity after the sale of the product, because it could be considered that the establishment is making improper use of Oluc or even selling it illegally. With regard to Article 6, the sections include:

§ Paragraph 1 - In order to comply with the obligation under the heading of this article, producers and importers may: I - hire a collection company duly authorized by the oil industry regulator; or II - qualify as a collection company, in accordance with the legislation of the oil industry regulator.

§ Paragraph 2 - Hiring a third-party collector does not exempt the producer or importer from responsibility for collecting and legally disposing of the used or contaminated oil collected. § Paragraph 3 - Producers and importers are jointly and severally liable for the actions and omissions of the collectors they hire (CONAMA, 2005, s/p).

In this way, the person responsible for marketing the products will have the greatest responsibility for complying with the Conama agreement, by hiring a specialized company to collect the Olucs.

According to ANP data, there are more than 300,000 outlets selling finished lubricating oils in Brazil. This, combined with the diversity of establishments such as auto parts, supermarkets, cooperatives and others, demonstrates the impracticality of environmental licensing for each one of them, but rather of establishing common rules for all of them, such as adopting the necessary measures to prevent Oluc from being mixed with other products (CONAMA, 2005).

Article 8: The Brazilian Institute for the Environment and Renewable Natural Resources (IBAMA), the oil industry regulatory body and the state environmental body, the latter, when requested, are responsible for monitoring and verifying that the collection percentages set by the Ministries of the Environment and Mines and Energy are being complied with.

Sole Paragraph: In order to carry out the control referred to in the heading of this article, Ibama will base itself on the information relating to the previous calendar quarter, and may therefore impose fines on establishments that do not comply with the law in force, as the legislation states with regard to costs and to whom such penalties are intended, in accordance with article 10:

The following finished lubricating oils are not included in the calculation basis for the amount of used or contaminated lubricating oil to be collected by the producer or importer:

I - for agricultural spraying;

II - for chainsaw chains;

III - industrial products that are part of the final product and do not generate waste;

IV - stamping;

V - for two-stroke engines;

VI - intended for use in sealed systems that do not require an oil change or involve a total loss of oil;

VII - soluble;

VIII - made from asphalt;

IX - intended for export, including those incorporated into machinery and equipment intended for export; and

X - all basic or finished lubricating oil sold between producing companies, between importing companies, or between producers and importers, duly authorized by the National Petroleum Agency - ANP (CONAMA, 2005, s/p).

In this way, Ibama acts as the body responsible for applying administrative penalties and fines for the incorrect disposal of Oluc, as there are still some environmental disasters committed by both large oil distribution companies and ordinary users of these products.

What does Art. 11: The Ministry of the Environment will maintain and coordinate a permanent monitoring group to follow up on this Resolution, which should meet at least quarterly, ensuring the participation of representatives of the oil industry regulator, producers and importers, retailers, collectors, re-refiners, entities representing state and municipal environmental bodies and non-governmental environmental organizations, thus making its applicability one of the most important articles of this resolution precisely the audit at these points where there is trade in Oluc. With the creation of this monitoring group, the audit on this issue of incorrect disposal will become intense, as contemplated below:

Art. 12: Any disposal of used or contaminated oils in soils, subsoils, inland waters, in the territorial sea, in the exclusive economic zone and in sewage or wastewater disposal systems is prohibited.

Art. 22: Failure to comply with the provisions of this Resolution shall result in the offenders being subject to, among others, the penalties provided for in Law No. 9.605, of February 12, 1998, and Decree No. 3.179, of September 21, 1999.

Art. 23: The obligations provided for in this Resolution are of relevant environmental interest.

Art. 24: Supervision of compliance with the obligations set out in this Resolution and application of the appropriate sanctions is the responsibility of IBAMA and the state and municipal environmental bodies, without prejudice to the proper competence of the oil industry regulatory body (CONAMA, 2005, s/p).

However, in Brazil, since 2005, Conama's Resolution 9 of 1993, which initiated the rules for recycling waste, has been repealed, and now Resolution 362/2005 has been adopted. The management of waste is a major concern for environmental organizations, as it can affect human health, damage soil and contaminate water basins, which is why care and auditing are required by law.

Federal Law 9.605/98 deals with environmental crimes of any order or degree, and discusses the penalties and administrative consequences of conduct and activities harmful to the environment. Article 54 states: "Causing pollution of any kind at levels that result or may result in damage to human health, or that cause the mortality of animals, or significant destruction of flora; Penalty: imprisonment, from one to four years and a fine." In any circumstance, whether in rural or urban areas or where there is a human grouping, this can result in administrative penalties and imprisonment depending on the case (BRASIL, 1998).

Similarly, there is Federal Law No. 9.847/1999, which determines the body responsible for overseeing the activities relating to the national fuel supply, which is the National Petroleum Agency (ANP), since Oluc is a petroleum derivative used illegally for fuel, and which provides for fines to be imposed in the event of infringements of the misuse of these derivatives (BRASIL, 1999).

In this way, the relevance of laws regulating the disposal of Olucs, storage and the reverse route, and the subjection of administrative penalties and sanctions imposed on both physical and legal persons in relation to environmental crimes imposed by the public authorities, becomes clear.

III. METHODOLOGICAL DESCRIPTION

In order to carry out this study, bibliographical research was carried out on the disposal of Oluc, legislation, reverse logistics, and the stages and procedures for recycling the products mentioned. Field research was also carried out on lubricant oil exchange points and how they are collected in the municipality of Guarapuava-Pr.

Taking into account the question of analysis and the objectives, we opted for quantitative research in which the report of this type of research is described using percentage data, graphs and tables. In order to outline the research, a case study was carried out, and finally the research results were analyzed.

The data collection instruments were interviews, bibliographical research on the subject and consultation of current legislation. The appendix contains an appropriate questionnaire from Rocha's work (2014), in places where new lubricating oils are sold and exchanged, in order to find an answer to the general objective proposed.

For this study, a questionnaire with 14 (fourteen) closed questions was constructed and the actors who work directly with the management of the Olucs were interviewed. We will therefore consider the interview as face-to-face, which according to Gil (2009) "has been considered to distinguish it from the questionnaire, whose items are presented in writing to the respondents".

Twenty (20) establishments in the city of Guarapuava that produce Olucs were chosen at random, in central and peripheral locations in the city, including six (6) mechanic workshops, six (6) fuel stations, three (3) auto centers, one (1) super oil change, one (1) transport company, one (1) light vehicle dealership, one (1) engine rectification and one (1) motorcycle workshop. Actors in the production chain considered to be the Oluc Generators, making it possible to obtain information related to the proposed objective.

3.1 The Case Study

According to Lüdke (1986), the case study is delimited by having characteristic contours whose interest meets the unit and does not make it evident.

How case studies are characterized:

1) Case studies are about discovery.
2) Case studies emphasize interpretation in context.
3) Case studies seek to portray reality in a complete and in-depth way.
4) Case studies use a variety of sources of information.

5) Case studies reveal vicarious experience and allow for naturalistic generalizations.

6) Case studies seek to represent the different and sometimes conflicting points of view present in a social situation.

7) Case study reports use a more accessible language and form than other research reports (LÜDKE, 1986, p. 18).

According to Yin (2010), data collection in a case study follows a formal protocol, but the specific information that may become relevant to the case study is not easily predictable. As information is collected from society, the case study becomes evident.

Following the same line of thought, the author states that good questions must be formulated and this will lead to mental and emotional exhaustion at the end of each day, but this exhaustion of analytical energy is very different from the experience of collecting experimental or survey data, i.e. testing subjects and administering questionnaires; in these situations, data collection is routine and the researcher must complete a certain amount of work.

According to Yin (2010), for case studies, "listening" means receiving information through multiple modalities. An example to be cited is making intense observations or sensing what may be happening in the research environment and not just using this modality. As an interviewee recounts an incident, the good listener hears the exact words used by the interviewee, picks up on certain behaviors and understands the context from which the interviewee perceives the world.

In this way, he suggests that the ability to listen is also important:

It needs to be applied to the inspection of documentary evidence as well as the observation of real-life situations. In document review, listening takes the form of worrying about whether there is any important message between the lines; any inference would of course need to be corroborated with other sources of information, but important *insights* can be gained in this way. Bad listeners may not even realize that there may be information between the lines; other deficiencies in listening include having a closed mind or simply having a bad memory (YIN, 2010, p.96).

Therefore, the case study is not a specific technique, as it aims to examine a particular environment, subject or situation in detail. The case study seeks to answer the questions "how", "why" and "when" certain phenomena occur, without having control over the events studied. This type of research is temporal because phenomena can be analyzed singularly in the current context, in real-life circumstances.

3.2 Analysis and Interpretation

The interviews took place with the people in charge of each establishment that generates

Olucs, and were carried out on site, with the aim of bringing the researcher closer so that none of the answers were hidden. The interviews were conducted in the form of notes on closed questionnaires, which can be found in the appendix. Some of the questions were designed to provide technical knowledge of the product as well as quantitative information, which will make the information pertinent to the adequacy of the analysis.

Analysis and interpretation vary significantly depending on the research plan, as the methodology for processing the data is the analysis and interpretation stage. Analysis will organize and compile the data in order to answer the research question. Interpretation seeks to make sense of this answer by linking it to the theories used in the framework (GIL, 2009).

The analysis will be based on the data collected via bibliographic research, including books, references to scientific articles, current legislation and the answers obtained in the interview, in order to visualize environmental responsibilities and an approach to sustainability in the management of Olucs.

IV. RESULTS AND DISCUSSION

This chapter presents the results and discussions relating to the specific objectives according to the study carried out in the city of Guarapuava-Pr on the management of Olucs, the correct storage for the disposal of Olucs, the investigation of the behavior of the lubricants market, and real compliance with the determinations imposed by environmental laws.

4.1 Geographical Scope of Olucs Collection

According to information from Sindirrefino (2012), the localities where Olucs are collected are covered. The collectors linked to this organization provide a regular collection service in 77% of Brazilian municipalities, with the North and Northeast regions having the most problems with this service, largely due to the distance from the country's main re-refineries, as can be seen in Table 1.

Table 1: Total number of municipalities collecting Olucs by region in Brazil

REGION	TOTAL MUNICIPALITIES	MUNICIPALITIES WITH REGULAR COLLECTION	PERCENTAGE OF MUNICIPALITIES ATTENDED
North	469	82	17%
North East	1830	1830	76%
Midwest	466	364	78%
South East	1668	1473	88%
South	1188	1012	85%
Brazil	6521	4328	77%

Source: SINDIRREFINO (2012).

It can be seen that 23% of Brazilian municipalities are not served by Olucs collection services, but the Southeast is the region where there is the greatest demand for these products and where the companies that re-refine Olucs are closest.

Still analyzing information from the same public entity, we obtained data from the southern region of Brazil, as shown in Table 2, which includes the municipality of Guarapuava, meeting the initial hypothesis regarding regular Olucs collections.

Table 2: Total number of municipalities collecting Olucs in southern Brazil

SOUTHERN REGION	TOTAL MUNICIPALITIES	MUNICIPALITIES WITH REGULAR COLLECTION	PERCENTAGE OF MUNICIPALITIES SERVED
Rio Grande do Sul	497	391	78%
Santa Catarina	295	244	82%
Parana	399	377	94%

Source: SINDIRREFINO (2012).

Among the states in the southern region, Paranà stands out, as it has the highest percentage of municipalities with a reverse logistics service that complies with current legislation. The city of Guarapuava in Paraná is covered by this service. These same results were found by Gusmao (2013) in his study of reverse logistics applied to lubricating oils in Boa Vista-RR.

It is crucial that the Oluc collection service is effective and carried out by a specialized company that complies with Conama's resolutions, thus avoiding obstacles in the Oluc production cycle, along with operational supervision by the National Petroleum Agency, which aims to manage the volumes of lubricating oil produced, imported and exempted from collection by state, as well as the volume of oil collected per collector (ANP, 2012).

Companies that collect used lubricating oil have various types of packaging available for storing it. It is up to the owner of the establishment, except for gas stations, to build underground tanks with double walls and continuous monitoring sensors to measure the amount of oil stored. Figure 4 shows the distribution of containers and their capacity in liters stored respectively: 20-liter drum, 200-liter drum and 1,000-liter container.

Figure 4: Types of packaging for storing Olucs Source: Sindirrefino (2012)

The establishment that carries out the oil change is the one that determines the packaging it will use to store the discarded oil and the destination it will receive, whether it is a collection company or the informal Olucs trade. In the case of Guarapuava, drums with a capacity of 200 liters are mostly used, where they are easy to handle when collecting Olucs.

4.2 The Collection and Disposal of Olucs in Guarapuava-Pr

According to data from the National Union of Fuel and Lubricant Distribution Companies (Sindicom), in 2015 it is estimated that around 400 million liters of used or contaminated lubricating oil (Oluc) were collected and sent for recycling, i.e. to the re-refining industry, representing a 52% increase on 2007. All of this waste was sent to re-refining units with a view to being reintroduced as an input into the lubricant production cycle, as proof of the full understanding of Sindicom's members of the concept of sustainability, as shown in Figure 5.

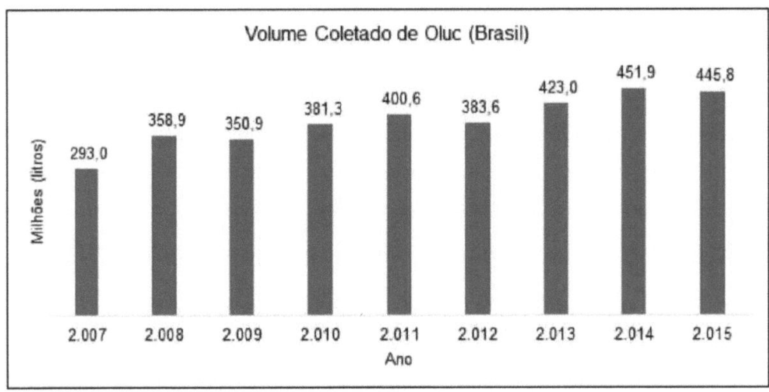

Figura 5: Volume Coletado de Oluc (Brasil)

Source: Sindicom (2016).

In the municipality of Guarapuava-Pr, there is no data on how much Oluc is collected, because there are several companies that work on the reverse route of these products, and the ANP does not publish information on cities, but on Brazilian regions and states.

In order to address the problem of this work, we investigated whether the establishments were aware of the environmental laws on recycling, and we noticed that some of them tried to do their best to comply with Conama's requirements. In some cases, they tried to carry out the Reverse Logistics of their products out of fear of committing offenses and incurring fines provided for by law, rather than out of awareness of sustainable development. This is evident from figure 6.

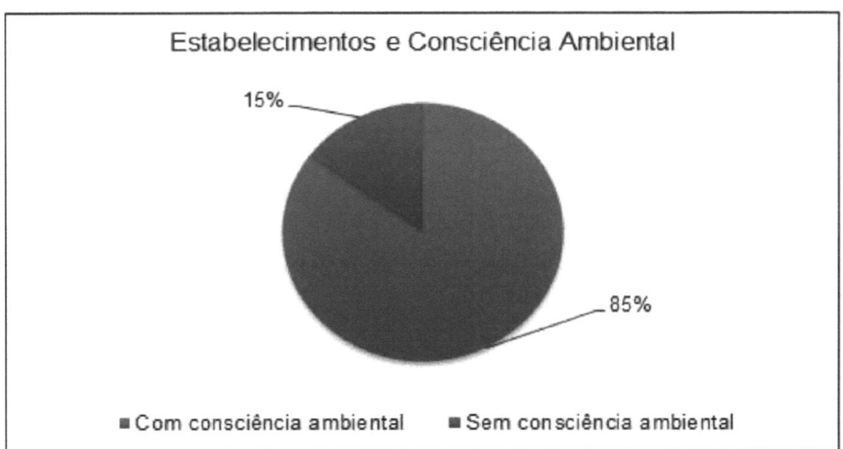

Figure 6: Establishments and Environmental Awareness Source: Field Research Results.

In the establishments where the interviews were carried out, there is effective Olucs collection in Guarapuava-Pr, mainly in central locations, in car centers, transport companies, grinding plants and fuel stations where there is more adequate supervision of the situation. In the more remote neighborhoods, there is a failure to collect the waste, mainly from mechanics' workshops, leading to incorrect storage, retention of the waste and also illegal trade, in violation of federal law 9.605/98.

In one of the mechanics' workshops that had problems with the management of the Olucs, the interviewee referred to the destination of the Olucs: *"Normally the customer himself asks me to take the "burnt oil" and in order not to lose the sale or the customer, I end up returning the product to him and I don't know what happens to it."*

In another mechanic's shop with the same problem located in a distant district of Guarapuava-Pr, the report was: *"There is a man who buys all the burnt oil to take to his farm, where he uses it to paint fences to combat pests that attack the wood and he uses it as fuel for his chainsaw".* It's a situation in which these businesses were never made aware of the rules to be followed and there was no technical visit for further explanations.

This practice must be reprimanded, as it goes against Conama's stipulations that all used or contaminated oil must be sent for re-refining, and this hidden dealing, which causes tax evasion, can have a negative impact on the local economy.

In this way, the interviewees make it clear that Conama's rules are not being complied with, and that they are unaware of the risk they are taking with regard to the clandestine disposal of the Olucs and the harm they are doing to society and to themselves.

In places where the Olucs are collected, the collection service usually comes to collect these products every 30 or 60 days, depending on the case and the demand for the product. The lubricating oil trade regularly undergoes seasonal changes, where sales increase and there is a need for the Olucs collection company to visit before the stipulated deadline.

Regarding the consumption of automotive lubricants in Guarapuava-Pr, there is no way of estimating demand, as there is no previous historical data, but what has been observed is the number of oil changes carried out monthly in each establishment, as shown in figure 7 below.

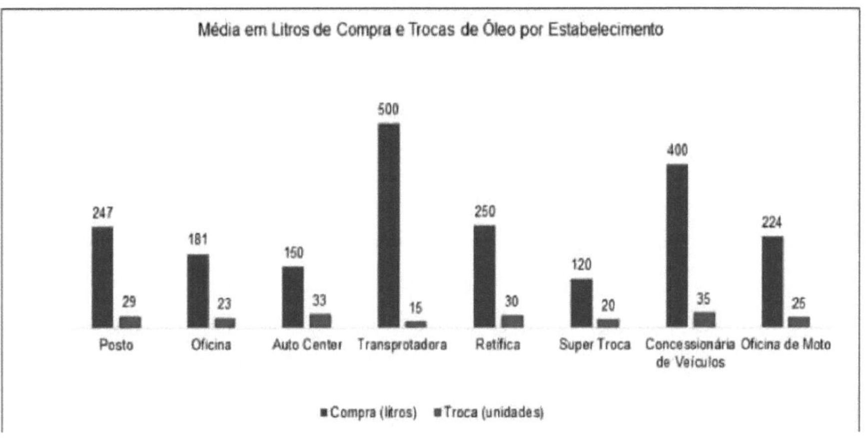

Figure 7: Average number of liters of oil purchased and exchanged per establishment Source: Field research results.

Figure 7 shows the purchases made by the sites visited from their respective suppliers of new lubricating oil, and their respective exchanges, which mean their sales, except for the transport company, which does not trade in these products, as it is considered a fleet operator, i.e. a final consumer. It is not possible to obtain an exact figure in liters of sales due to the varying capacity of the crankcase[2] of each motor vehicle.

According to the sites visited, there is a demand for lubricant brands, due to market recognition, partnerships with dealers, better prices or the technology used to manufacture the lubricant.

With regard to the brands preferred by generators at the time of purchase, it can be seen that the Petrobrâs brand is more widely accepted in Guarapuava, due to brand recognition,

2 Meaning of Carter: One of the parts of the engine where the oil is stored, responsible for lubricating the engine.

because it is of Brazilian origin and has a competitive price compared to the others. The Motul brand is used by only one of the interviewees, due to its manufacturing technology, product quality and the fact that the price is not relevant to end consumers.

Figure 8 shows the volume in liters purchased each month, and the respective brands by segment

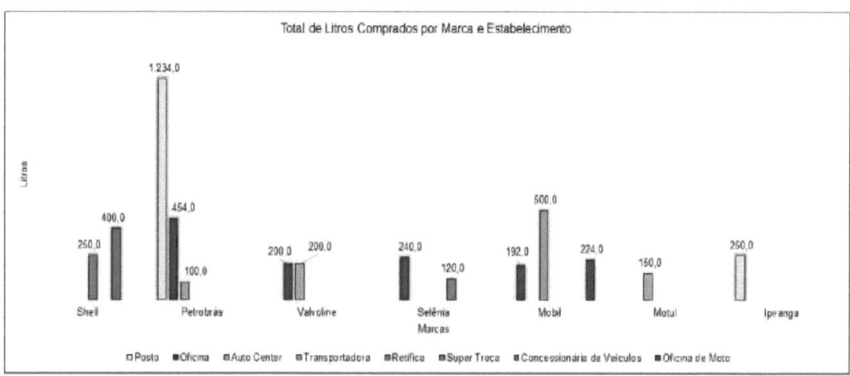

Figure 8: Total Liters Purchased by Brand and Establishment

Source: Field research results.

In this case, it is up to the final consumer to choose the lubricant of their choice. Choice seeks to describe how consumers make purchasing decisions and how they face tradeoffs[3] and changes in their environment (PINDYCK, 1994).

Due to the exposure of workers to the dangers of contamination via Olucs, there must be an effort to use Personal Protective Equipment (PPE), as this is an item of fundamental importance in preventing possible accidents, chronic non-communicable diseases such as diabetes and hypertension, because in these companies there must be a model for preventing accidents and occupational diseases that combines the simplicity of the intervention and the depth of the technical actions needed for it to be effective (GARDINALLI, 2012).

Among the facilities checked on site by the researcher, it was noted that there is no management manual for Olucs anywhere, and that oil changes are mostly carried out in elevators due to the lower cost to the company. It was found that the operators who carry out the oil changes do not use any kind of personal protective equipment, such as PVC

3Meaning of Tradeoff : is an English expression meaning the act of choosing one thing over another, implying a conflict of choice and a consequent relationship of compromise.

gloves, rubber boots, aprons or goggles to protect their eyes, and consequently there is no emergency plan in case of accidents. Except in car dealerships, where there is a frequent audit by the car industry.

V. FINAL CONSIDERATIONS

The aim of this work was to evaluate the Olucs cycle chain in Guarapuava-Pr, through the actors involved in the process, how Reverse Logistics takes place and the laws that guide this system.

Environmental awareness can be awakened by choosing establishments that dispose of lubricating oils in accordance with current legislation, thus highlighting the importance and value of the environment in which we live, since the issue of environmental sustainability has become something of a buzzword in society.

Because the researcher has technical and application knowledge of automotive lubricating oils, it motivated him to bring something scientific, in order to corroborate with pertinent information to the readers of this research so that they also have the habit of conserving the environment.

Regulated environmental laws make it clear how necessary it is to manage Olucs well so that everyone wins, especially in this sector of the economy, thus reducing the environmental impact caused by those who don't comply with their requirements.

One of the main difficulties is to take care of and control the correct disposal of oil, even though it is a petroleum product that is considered to be exhaustible, there is no prospect of reducing its production and importing it, because this is one of the sectors that is developing the most, along with the growth of the car fleet in Brazil.

In this sense of taking care with disposal, there needs to be more government interference with some of the players in this production chain, in this case the retailers, so that they don't sell their products to anyone who isn't committed to complying with their environmental obligations.

One of the most important things that can be done is for the bodies responsible for this environmental segment to carry out more inspections and audits, but not only in central locations where oil changes are common, but also where they are not frequent or where there is little flow of people.

The responsible public administration and the private sector would have to mobilize the community on issues of sustainable development, in order to boost knowledge of the management of Olucs, via lectures, training and a greater number of technical visits to answer questions from those who work with the handling of these products.

This shows that in Guarapuava-Pr, there is a correct cycle of waste in some establishments, where Reverse Logistics plays its part. On the other hand, in some places Conama's requirements are not being met, contrary to the initial hypothesis mentioned. The lack of knowledge on the part of those involved in the process is significant, as is the lack of supervision by the government body responsible for this line of work.

In view of the results found, it is suggested that informative tools be created, such as folders and information about the correct disposal of Olucs in all establishments that work with Olucs, as it was found that there is no information about the harm that can be caused by the incorrect disposal of Olucs.

The findings of this study show that Guarapuava-Pr lacks an Olucs collection center, due to its location and junction, as it is the largest city in the South Central region of the state. This venture could encourage greater Olucs collection and more efficient service to collection points.

The questions raised in the course of the research and the solutions found are coherent, leading us to reflect on the proper collection of Used Lubricating Oils, and for future research we recommend further investigation into recycling programs and environmental policies based on the management of Olucs.

VI. REFERENCES

ABETRE, Brazilian Association of Waste Treatment Companies. ***Classification of Solid Waste ABNT Standard NBR 10.004:2004.*** Available at: http://www.abetre.org.br/biblioteca/publicacoes/publicacoes-abetre/classificacao- de-residuos Accessed on November 6, 2016.

BNDES. National Development Bank. ***Diversification potential of the Brazilian chemical industry. Report 3- Lubricating oils.*** Available at :http://www.bndes.gov.br/SiteBNDES/export/sites/default/bndes en/Galerias/Arquivos /produtos/download/aep fep/chamada publica FEPprospec0311 Lubrificantes.pdf Accessed on April 23, 2016.

BRAZIL. Law No. 9.605, of February 12, 1998. ***Provides for criminal and administrative sanctions arising from conduct and activities harmful to the environment, and makes other provisions.*** Official Gazette of the Union, Brasilia, DF, February 13, 1998.

BRAZIL. Law No. 9847, of October 26, 1999. ***Provides for the inspection of activities relating to the national supply of fuels, referred to in Law No. 9.478, of August 6, 1997, establishes administrative sanctions and makes other provisions.*** Official Gazette of the Union, Brasilia, DF, October 27, 1999.

BRAZIL. Law No. 12305, of August 2, 2010. ***Institutes the National Solid Waste Policy;*** amends Law No. 9605, of February 12, 1998; and makes other provisions. Official Gazette of the Union, Brasilia, DF, August 3, 2010.

Concepçoes da economia ecológica: suas relações com a economia dominante e a economia ambiental. Estudos avançados, vol. 24, n° 68. Sâo Paulo, 2010. Available at: http://www.scielo.br/scielo.php?script=sci arttext&pid=S0103-

40142010000100007&lng=pt&nrm=iso. Accessed on: April 16, 2016.

CONAMA. ***National Environment Council.*** Resolution No. 362/2005, of June 27, 2005. Provides for the collection, disposal and final destination of used or contaminated lubricating oil. Available at: http://www.mma.gov.br/port/conama/legiabre.cfm?codlegi=466. Accessed on: January 15, 2016.

. ***Used or Contaminated Lubricating Oils***: Guidelines for environmental licensing. CONAMA Resolution 362/2005. Available at: http://www.mma.gov.br/port/conama/processos/6ACA4025/Manual orientacao.p df. Accessed on: March 15, 2016.

FAUCHEUX, S., NOËL, J.F..***Economia dos Recursos Naturais e do Meio Ambiente***. Lisbon-Portugal: Instituto Piaget, 1995.

GARDINALLI, J. R.. ***Manual for the Prevention of Accidents and Illnesses at Work.*** Available at : http://www.trajanocamargo.com.br/wp- content/uploads/2012/05/seguranca no trabalho.pdf Accessed on October 3, 2016.

GIL, A. C..***Métodos e técnicas de pesquisa social***. 6ª ed., Sao Paulo: Atlas, 2009.

GUSMÂO, J.G.S. ***REVERSE LOGISTICS APPLIED TO USED OR CONTAMINATED LUBRICANT OILS PRODUCED AT FUEL STATIONS IN THE CITY OF BOA VISTA-RR.*** Available at: http://200.230.184.11/ojs/index.php/CCHAS/article/view/20/10 Accessed on August 22, 2016.

GONÇALVES,R.A.O.. ***The Importance of Rerefining Lubricating Oils***. Available at: ***http://techoje.com.br/site/techoje/categoria/detalhe*** article/1180Accessed on March 19, 2016. IBAMA. ***Brazilian Institute for the Environment and Renewable Natural Resources***. Available at: http://ibama.gov.br/acesso-a-informacao/atribuicoes. Accessed on March 21, 2016.

LEITE, P.R..***Logistica Reversa: Meio Ambiente e Competitividade***. Sao Paulo: Pearson Prentice Hall, 2009 2nd Edition.

LOYOLA G, R..***Environmental Economics and Ecological Economics: A Theoretical Discussion***. II National Meeting of ECOECO. Sao Paulo, 1997. Available at: http://www.ecoeco.org.br/publicacoes/encontros/106-ii-encontro-nacional-da-ecoeco- sao-paulo-sp-1997. Accessed on: February 16, 2016.

LÜDKE, M. ***Pesquisa em educação: abordagens qualitativas***.Sao Paulo: EPU, 1986.

MAY, P. H..***Economia do Meio Ambiente: teoria e pràtica***. 2nd ed., Rio de Janeiro: Elsevier, 2010.

MOURA, L. A. A..***Economia Ambiental: Gestão de Custos e Investimentos***. 3rd ed., Sao Paulo: Editora Juarez de Oliveira, 2006.

PEREIRA, A.L. et al .***Reverse Logistics and Sustainability***. Sao Paulo: Cengage Learning, 2013.

PINDYCK, Robert S. RUBINFELD, Daniel L. ***Microeconomia***. Sao Paulo: Makron Books, 1994.

ROCHA, B. S.; SCALIZE, P.S.; ARRUDA, P.N; ET AL. ***Management of used lubricant oil at fuel stations in the municipality of Terezòpolis de Goiàs - GO,*** **Brazil.** Available at: https://periodicos.ufsm.br/remoa/article/viewFile/15193/pdf. Accessed on July 23, 2016

SILVA, L. F. S.; SOUZA, G. S.; SILVA, T. T.; ET AL. ***Oil: oil spills and their impact on the environment.***Simpósio Internacional de Ciências Integradas da Unaerp Campus Guarujà, 2009. Available at: http://www.unaerp.br/sici- unaerp/edicoes-anteriores/2009/secao-1-5/1087-petroleo-derramamento-de-oleo-e- seus-impactos-no-meio-ambiente/file. Accessed on: April 16, 2016.

SINDICOM. ***National Union of Fuel and Lubricant Distribution Companies.*** Available at:

http://www.sindicom.com.br/#conteudo.asp?conteudo=97&id pai=104&targetElement=leftpart. Accessed on September 08, 2016.

SINDIRREFINO. ***National Union of the Mineral Oil Re-refining Industry.*** Available at: http://www.sindirrefino.org.br/coleta/centros-de-coleta.

Accessed on: March 19, 2016.

SOHN, H..***Disposal of used lubricating oil: environmental and social aspects.*** Available at:

http://www.mma.gov.br/port/conama/reuniao/dir857/Hassan.pdf. Accessed on: April 4, 2016.

TRISTAO,J.A.M..***Environmental Management of Lubricating Oil Waste: the***

Re-refining process. Available at :http://www.anpad.org.br/diversos/trabalhos/EnANPAD/enanpad 2005/APS/2005 APSC2161.pdf Accessed on May 04, 2016.

YIN, R.K.: ***Case study Planning and methods***. Porto alegre, Bookman, 4 ed,2010.

APPENDIX 1: FIELD QUESTIONNAIRE

1) Is the company aware of the laws and recycling of Used Lubricating Oils? ( ) Yes ( ) No

2) Does the company account for the volume of Lubricating Oil? ( ) Yes ( )No

3) Which brand do you prefer? ____________________________________

4) Does the company send Oluc for re-refining? ( ) Yes ( ) No

5) What is the interval between collections by the company responsible? ( ) 30 days ( ) 60 days ( ) 90 days or more

6) Oil changes are carried out in: ( ) Box ( ) Elevator ( ) Area with waterproof floor and drainage system

7) Oluc is stored through: ( ) Underground tank of double-walled with monitoring sensors ( ) Drums, Pump, Containers ( ) Other storage system

8) How many oil changes do you do each month? () 10 to 20 ( ) 21 to 49 ( ) 50 or more

9) Is there an emergency plan in case of accidents involving Oluc? ( ) Yes ( ) No

10) Is information on the management and generation of Oluc available to end consumers? ( ) Yes ( ) No

11) Is there an audit of the destination of the Oluc by the responsible government body? ( ) Yes ( ) No

12) Is there a manual adopted by the establishment for the management of Oluc? ( ) Yes ( ) No

13) Is a certificate issued at the time of collection? ( ) Yes ( ) No

14) Does the company use PPE (Personal Protective Equipment - PVC gloves, rubber boots, apron or goggles) during oil change activities? ( ) Yes ( ) No

Printed by Books on Demand GmbH, Norderstedt / Germany